Interior Design Review

室内设计奥斯卡奖

安德鲁·马丁国际室内设计年度大奖
2012/2013获奖作品

[英] 马丁·沃勒（Martin Waller） 编著
北京安德马丁文化传播有限公司 总策划
《FRAME国际中文版》杂志社 出版策划

前 言

为长时间汽车旅行而保留的一个著名的毫无意义的游戏就是给有史以来最伟大的演员、运动员和政治家评等级（詹姆斯·斯图尔特James Stewart、穆罕默德·阿里Mohammed Ali和温斯顿·丘吉尔Winston Churchill）。对比来说给室内设计师评等级同样不公平，纵贯整个历史来评等级则尤其荒谬。然而，以下是安德鲁·马丁评选出的有史以来十位最伟大的室内设计师，他们按逆序排列依次是：

十、比利·鲍德温（Billy Baldwin, 1903-1983），"美国装饰界的教父"，定义了美国战后的审美，为美国杰出的室内设计定下了基调。

九、托马斯·杰斐逊（Thomas Jefferson, 1743-1826），唯一一个同时在演员、运动员和政治家的辩论中起重要作用的人。第三届美利坚合众国总统巩固了联邦风格的影响力，他在蒙蒂塞洛（Monticello）的多面设计仍然是美国室内设计发展的转折点。他还起草了美国《独立宣言》（the Declaration of Independence）。

八、勒·柯布西（Le Corbusier, 1887-1965），尽管他的政治是狡猾的，有时候建筑是不可靠的，但是勒·柯布西的"居住机器"（machine for living）理论影响了整个20世纪。就像在1930年一样，他设计中使用的白墙、裸露的灯泡和有管状框架的家具构建出一种对现今依然存在重要作用的纯粹性。

七、艾尔西·德·沃尔夫（Elsie de Wolfe, 1865-1950），《纽约客》（New Yorker）杂志声称："室内设计作为一个专业是被艾尔西·德·沃尔夫创造的。"1913年她出版了《品位饰家》（The House in Good Taste），书中提出了一个明亮、清新、有女性活力的领域。就像她说的："我打开了美国的门窗，让空气和阳光得以进来"。

六、罗伯特·亚当（Robert Adam, 1728-1792），是英国乔治王朝时期最有影响力的建筑师、室内设计师和家具设计师，以他的名字命名的"古典洛可可式"风格的灯具被认为是英式品味的极致（尽管他是苏格兰人）。罗伯特·亚当被埋葬在威斯敏斯特教堂（Westminster Abbey），他的护柩者包括三个伯爵和一个公爵。

五、大卫·希克斯（David Hicks, 1929-1998），是典型的豪华风格室内装饰师，大卫·希克斯是魅力、绅士和成熟教养的完美模范，是一个尽善尽美的天才。他高雅的精确度和低调的华丽为大批的追随者设定了一个基准。

四、查尔斯·勒·布伦（Charles Le Brun 1619-1690），被法国路易十四国王（Louis XIV）称为有史以来最伟大的法国艺术家，杰出的勒·布伦同时也是凡尔赛宫（Versailles）的主要室内装饰师。他在1660年成立了生产各式各样家具的法国巴黎哥白林挂毯厂（随后Gobelins成了挂毯的一个俗称）。有人说："在王国制造的任何东西都是建立在勒·布伦的素描和绘画的基础上。"因此他几乎以一人之力创造了路易十四时期的语言。

三、刘秉忠（Liu Bingzhong, 1216-1274），当马可·波罗（Marco Polo）在元上都（Xanadu）皇宫朝见元世祖忽必烈（Kublai Khan）时感到大为吃惊，他对元上都非凡风采的记录激发了至今已经持续700年的中国艺术风格（Chinoiserie）。曾经是僧人的刘秉忠是这整座城市的设计策划者，他被认为拥有神秘的力量和预示未来的能力，他的工作成果将在塞缪尔·泰勒·柯勒律治（Samuel Taylor Coleridge）的诗篇中永垂不朽。

二、米开朗基罗（Michelangelo, 1475-1564），不仅仅是雕刻家、画家、诗人和工程师，同时也是建筑师和室内设计师，他的室内设计最好的例子是劳伦齐阿纳图书馆（Laurentian Library）。米开朗基罗对接收的想法进行彻底地重新整理，促使了对安德烈亚·帕拉第奥（Andrea Palladio）有极大影响的矫饰主义（Mannerist）运动的产生。他在74岁的时候设计了圣彼得大教堂（St. Peter's Basilica）。在你需要为天花板绘画或者为前厅设置雕像的时候他同样能帮助你。

一、菲狄亚斯（Phidias, 490BC-430BC），菲狄亚斯执导了帕特农神庙（Parthenon）的建设，同时负责了其中巨型雅典娜（Athena）雕像的雕刻工作。然而，菲狄亚斯不仅是远古世纪最伟大的雕刻家，也是比任何人都要早的古典希腊风格的推动者，该风格在2500年来的西方文化中始终占据主导地位。此外，他在奥林匹亚（Olympia）创造了世界七大奇迹之一的宙斯神像（The Statue of Zeus），神像高12.8米（42英尺），由黄金和象牙制造而成。菲狄亚斯被诬陷盗窃，最后在监狱中身亡。

马丁·沃勒（Martin Waller）

Geoffrey Bradfield

Designer: Geoffrey Bradfield. Company: Geoffrey Bradfield Design, New York, USA. Specialising in luxurious residences and offices for an international clientele. Current projects include a 100,000 sq ft residence in Central Tokyo, a 300,000 sq ft estate outside Shanghai and a penthouse in Jerusalem with three apartments below for the residents' children. Recent work includes a 7,000 sq ft duplex penthouse in Mexico City, the subject of newly released 'A 21st Century Palace', the New York apartment of Hollywood Director, Oliver Stone, facing the Hudson and a 6,000 sq ft apartment for a Russian family on Madison Avenue, New York City.

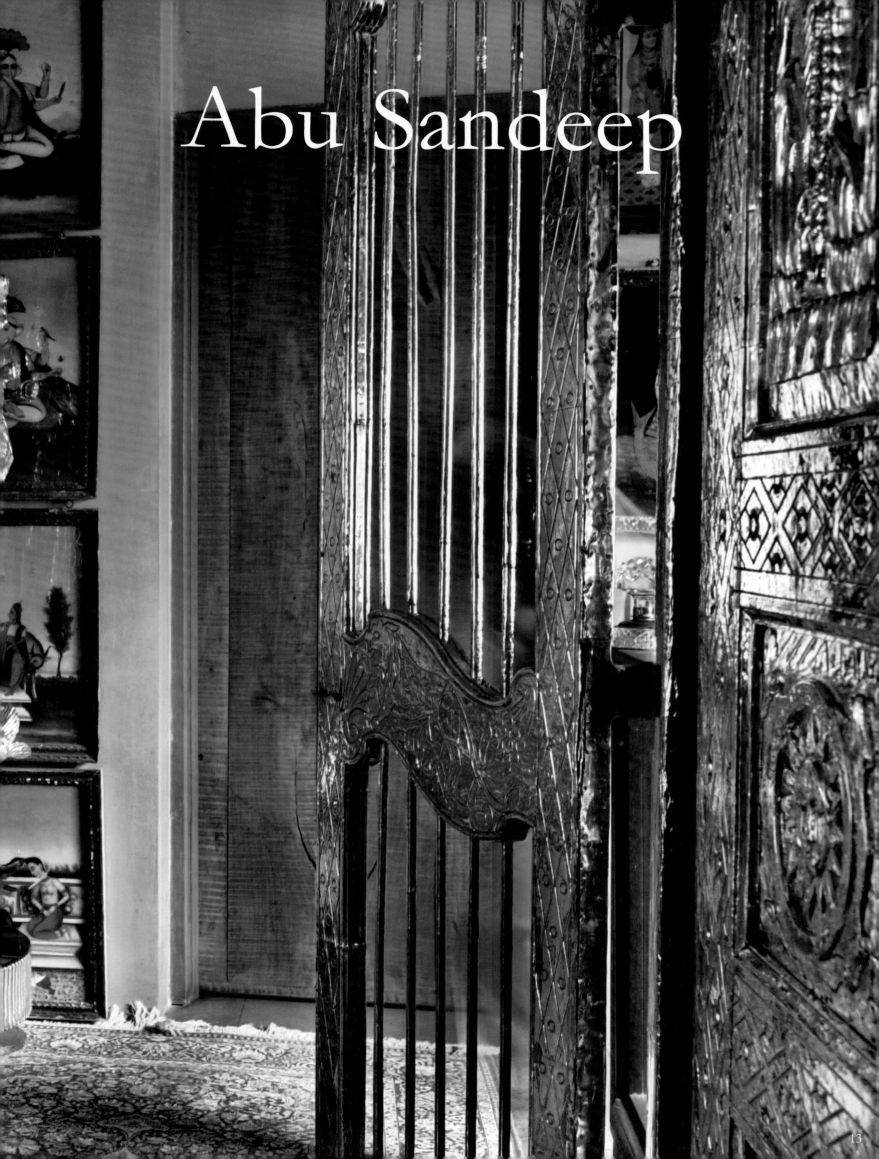

Abu Sandeep

Designers: Abu Jani and Sandeep Khosla. Company: Abu Jani Sandeep Khosla, Mumbai, India. The country's premier couturiers who diversified from their core business of fashion to enter the world of Interior Design in the mid 90s. Specialising in residential projects, they have designed homes in London, Mumbai, Hrishikesh and New Delhi. Recent work includes Mohinder Verma's New Delhi residence, The Passi house in New Delhi and a villa in the hills of Hrishikesh, North India. Their signature style is maximalism and attention to detail, their skill is combining art, artefacts and antiques to create homes fit for modern day Maharajahs.

Jorge Cañete

Jorge Cañete Interior Design Philosophy, Saint - Legier, Switzerland. At the Maison et Objet fair in Paris 2012, The International Interior Design Association presented Jorge Cañete and his studio the Global Excellence Award for best international project in the residence category. His latest works include: Isa Barbier contemporary art scenography in an 18th century castle, various major residential projects in Switzerland and apartments, villas and castles in Europe. 'Cañete is distinguished by a style that seeks to surprise at the same time as it offers a poetic vision of the world'.

Juan Montoya

Company: Juan Montoya Design Corporation, New York. Columbian born Juan studied architecture in Bogotá then moved to New York where he graduated from Parsons School of Design. Following two years of work and study in Paris and Milan he returned to New York where he founded his company. Specialising in residential and contract interior design with projects located throughout the United States and Internationally. He also designs and markets an extensive line of furniture for Century Furniture. Recent work includes apartments in both NYC and Paris and a mansion in Moscow. Current projects include a gallery in Paris, Galerie Agnes Monplaisir, a family compound in Mexico and a house in Cap Cana, Dominican Republic. Design philosophy: 'Give me your dreams and I will make them happen'.

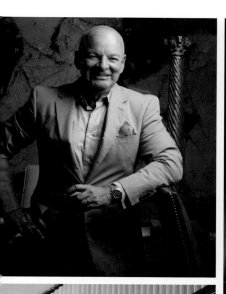

Hare+Klein

Designer: Meryl Hare. Company: Hare+Klein, Sydney, Australia. An award winning practice founded in 1988. Hare+Klein specialise in residential and commercial projects worldwide. Recent work includes a luxury boutique island resort in Fiji, a resort on Whitsunday Islands and various homes in Sydney. Current projects include an historical home in Adelaide, executive boutique offices in Sydney, a restaurant on Hamilton Island and a beach house in Palm Beach. Design philosophy: committed to creating original and individual interiors of quality that stand the test of time.

David Rockwell

Designer: David Rockwell. Company: Rockwell Group, New York, USA. Long before turning his attention to architecture, David Rockwell was a child of the theatre, his mother worked as a vaudeville dancer and cast him in repertory productions. This brought David's passion for spectacle and his eye for colour to his architecture training. Now founder and CEO of Rockwell Group, he brings that sensibility to a diverse array of projects from hotels (W and YOTEL) to cultural institutions and theatres (Elinor Bunin Munroe Film Center at Lincoln Center, Kodak Theatre in Los Angeles), playgrounds (Imagination Playground),

restaurants airport terminals (JFK's Jet Blue), Broadway sets (Hairspray) and more. Current work includes Nobu Hotel, Las Vegas, NV and a furniture collection with Moroso.

Joseph Karam

Designer: Joseph Karam. Company: Galerie Joseph Karam, Paris. Lebanese born Joseph devoted his time to studying interior design before settling in France. Thirty years on, he has expanded to Beirut and Moscow and built up a team of 40 experienced architects, interior designers and furniture designers. The luxurious and refined interior he creates meets the expectations of a demanding international clientele. Throughout the years, his diverse creative approach has helped him develop his specific sense of style that now defines his position in the world of interior design.

Heidrun Diekmann

Designer: Heidrun Diekmann. Company: Heidrun Diekmann Interior Lifestyles CC, Windhoek, Namibia. A small team specialising in a range of high end residential and commercial projects. Heidrun is currently working with a Paris based client to rebuild his game lodge in Namibia, a private estate in Dakar and a new build family home in Namibia. Recent projects include a farmhouse in Africa, a family mansion in Windhoek as well as a lodge situated in the vicinity of the Etosha Pan of Namibia. Design philosophy: to create soulful spaces with a sense of peace and tranquility.

One Plus Partnership

Designers: Ajax Law and Virginia Lung. Company: One Plus Partnership Ltd, Hong Kong. Established in 2004, One Plus initiates each project by a specific theme and develops it into a distinct space. Current work includes a coffee shop, sales office, clubhouse and a 57,000 sq ft cinema and children's cinema. Philosophy: to deliver unconventional and beautiful design.

Judy Hutson

Designer: Judy Hutson. Since 1994 Judy has worked as a designer specialising in hotels and restaurants. Projects include The Pig Hotel in Brockenhurst, Hampshire, a fresh take on the traditional country house hotel with an eclectic mix of old and new giving the interiors an evolved style. The ethos of The Pig is driven by its walled garden that not only provides much of the produce for the restaurant but design clues for the entire project. Current work includes The Pig in the Wall, Southampton, the refurbishment of a 5 star luxury boutique hotel in the New Forest, Wayneflete House a private residential property, Hotel Crillon Le Brave, Provence and various hotels in the UK for the Hotel du Vin group.

Jeffers Design Group, California, USA. Recognised as one of the most dynamic designers of his generation, through his urbane use of colour and pattern Jeffers has established a cool, collected style that is very much his own. Recent projects range from a classic Manhattan residence to an historic San Francisco estate to a mid century modern-inspired home in Los Angeles. In 2011, Elle Decor named Jeffers to a coveted position on its A-List of the country's top 25 interior designers and he has been widely published in the U.S. and internationally. Jeffers launched his first retail venture, Cavalier, in San Francisco in 2012. His design philosophy is inspired by Billy Baldwin's mantra 'Interior design is the art of arranging beautiful things, comfortably.'

Silvio Rech and

Lesley Carstens

Company: Silvio Rech & Lesley Carstens Architecture and Interior Architecture, Westcliff, South Africa. A husband and wife team who over the past three decades have redefined the contemporary architectural landscape in South Africa. Recent projects include 'Randlords', an exclusive rooftop bar in the heart of urban Johannesburg, Villa 11, North Island, Seychelles, the honeymoon location chosen by the recently married Duke and Duchess of Cambridge, The Waterberg Telescope Observatory and the award winning Rammed-Earth Pavilion in Westcliff Johannesburg. This work represents a further move for the practice towards more contemporary, sculptural forms.

Sam Cardella

Cardella Design, California, USA. Projects are residential, with an emphasis on new architectural design and renovations. Recent work includes a desert golf course estate renovation, a Baltimore high rise condominium and a Malibu Colony beach house. Current projects include an Albert Frey architectural restoration in Palm Springs, a private ranch estate and several renovations in Palm Springs, Rancho Mirage and Palm Desert, California, as well as a Chicago condominium residence on Miracle Mile. Cardella embraces his clients' personality and lifestyle, allowing the architecture to direct the overall design.

Shi Xudong

Designer: Shi Xudong. Company: Xuridongsheng, Fujian, China. Work is predominantly in China. Current projects include Fujian Fuzhou Cultural Palace Club, Alxa League Tianyi Group Sales Department and Inner Mongolia Tongliao Tea Club design. Recent work includes Wuyishan tourism experts floor design, Fujian Training Hotel and Hefei swallow's Nest Restaurant. Design philosophy: traditional oriental elements for inspiration.

Designer: Kirk Nix. Company KNA Design, California, USA. With 25 years experience across many of the world's most distinctive properties, KNA specialise in original, timeless and functional work within the hospitality industry. Recent projects include The Venetian Casino and Resort, The Hotel Wilshire, William Morris/ Endeavor. Current work includes The Caesar's Palace Octavius Villas, The Disney Hotel-Shanghai Resort and The Conrad Hotel-Guangzhou. Design Philosophy: to find the perfect blend of humour and sophistication in every interior.

KNA Design

111

Graça Viterbo

Viterbo Interior Design, Estoril, Portugal plus offices in Angola & Singapore. Internationally awarded practice founded by Graça Viterbo in 1971. From 2000 it became a mother and daughter partnership with a 27 staff team. Projects are tailor made and both private and commercial, including numerous luxury and boutique hotels. Recent work includes Bela Vista Hotel and Spa, credited by The New York Times as one of the must go places in 2012. From designing windows for Hermès to publishing her latest book LIFEstyle, as well as currently designing homes in Bangkok, Singapore, Switzerland, Luanda, Lisbon and Brazil, Gracinha follows in her mother's footsteps.

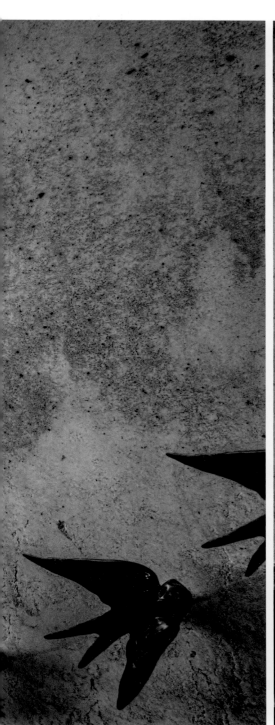

Matrix

Designers: Wang Guan, Liu Jianhui, Wang Zhaobao. Company: Shenzhen Matrix Design Co, Ltd, China. Founded in 2010 with a team of young and prominent designers. Recent work includes Hua Di Zi Yuan Show Flat in Anhui Province, Chongqing Vanke Office and Tian Yue Wan Villa Show Flat in Shenzhen. Current projects include Zi Peng Mountain Vacation Clubs in Anhui Province. Matrix's design philosophy is to create a new world in interior design.

Designers: Bunny Turner & Emma Pocock. Company: Turner Pocock Interior Design, London. Specialising in high end residential work and boutique commercial projects. Recent work includes the full design of Chelsea restaurant 'Medlar'. Current projects include consultant design for the redevelopment of three luxury flats and the complete reconfiguration of a Chelsea Mews house into an open plan loft style living space.

Turner Pocock

Marc Hertrich & Nicolas Adnet

Studio Marc Hertrich & Nicolas Adnet, Paris, France. Predominantly residential work including spas, restaurants, hotels and resorts. Recent projects include prestige suites and rooms at Hotel Martinez, Cannes, restaurant 'L'Instant d'Or', Paris, boutique hotel 'Les Jardins de la Villa', Paris, a mountain residence in Crans-Montana, chalets and Club Med 'Valmorel' resort in the French Alpes and 'Moofushi' resort in the Maldives. Current work includes a hotel in Casablanca, new villas in Marrakech, a resort in the region of Guilin, China and a gastronomic restaurant on a golf course. Design Philosophy: create luxurious, comfortable and elegant interiors with individuality and surprise.

Elena Akimova

Elena Akimova Design, Moscow, Russia. Predominantly residential work with some commercial throughout Moscow and Europe. Recent projects include country houses in Moscow, the South of France and Spain plus apartments in Moscow and Vienna as well as the interior design of a private jet. Current work has been carried out with the help of architect Katya Andreeva.

Brian Gluckstein

Gluckstein Design Planning Inc. Ontario, Canada. Specialises in interior design, planning and project management services for prestigious residential and corporate clients, residential developers and the hospitality and leisure industries. Recent work includes the Spa at Four Seasons in Palm Beach, Four Seasons Toronto and an Aspen ski chalet. Current projects include a Toronto penthouse, a beach front house in Palm Beach and a Toronto loft. Brian's design philosophy is to create beautiful interiors that are sophisticated and luxurious as well as comfortable and effortless.

Daun Curry

Modern Declaration, New York, USA. A boutique design firm with a collection of custom furniture and lighting, offering a full design service for high end residential and commercial clients. Current work includes a Soho showroom

for accessory designers Stella & Dot, a New York pied-à-terre for London business tycoon Pierre Lagrange featuring a notable art collection, marketing offices and product academy for luxury beauty brand Moroccanoil in New York, a five bedroom family flat for Moroccanoil co-founder Carmen Tal, the complete renovation of a townhouse in NYC's West Village, complete with private screening room, solarium, wine vault and an outdoor oasis complete with organic vegetable garden. Design philosophy: Be brave and make a bold statement.

Feng Yu

Designer: Feng Yu. Company: Deve Build Design, Shenzhen, China. Recent projects include a private club 'The Oriental Club', Shenzhen, The Tianxi Oriental Club, Huizhou and an apartment in Huizhou. Current work includes a hotel and a villa in Huizhou plus various clubs and apartments. Design philosophy: engineering meets design.

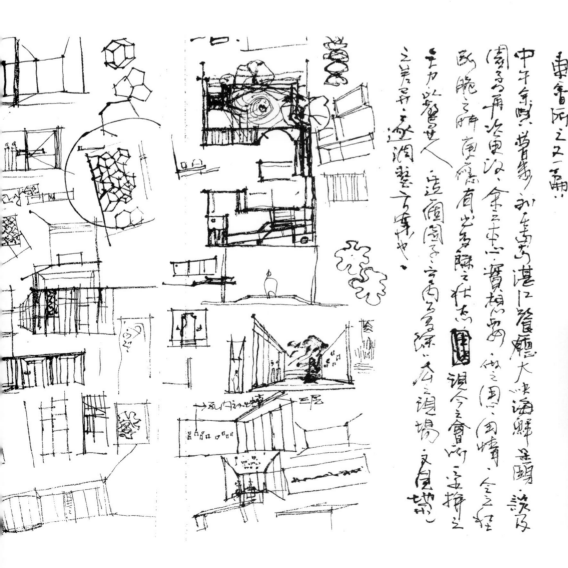

Casa do Passadico

Designers: Catarina Rosas, Cláudia Soares Pereira and Catarina Soares Pereira. Company: Casa do Passadiço, Portugal. Bespoke residential projects. Current work includes the fittings of a 40 metre private yacht, a contemporary private house in Estoril facing the sea, a modern villa in Vale do Lobo, Algarve, a private villa in Luanda, Angola, a pied-à-terre in a Hôtel Parculier, Saint-Germain-des-Prés, Paris as well as several private houses in Portugal and abroad. Their design philosophy is to create quality, that's sophisticated, luxurious, comfortable and balanced.

Julia Buckingham Edelmann

Designer: Julia Buckingham Edelmann. Company: Buckingham Interiors + Design, Chicago, Illinois, USA. Current work includes an historic Victorian Graystone Mansion plus coach house conversion in Lincoln Park, the full refurbishment of 6 flats into a single family home with spa and rooftop garden in Chicago and a Victorian mansion in Northern Iowa. Recent projects include a North Shore Home, a Michigan Avenue pied-à-terre, a Nantucket restoration and the redesign of a North Shore French Estate. Design philosophy: to incorporate a mixture of styles and eras with antiques and artifacts.

Bo Li

Designer: Bo Li. Company: Cimax Design Engineering, Hong Kong. Having joined the company in 2004, Li heads a young team with Jie Wen to provide a broad design service for its clients. Current projects include designing an office and restaurant of a club that was transformed from an old plant in Beijing and the architecture and interior design of a 3,000 sq m private villa in Shanghai. Recent work includes a 1,000 sq m golf villa in Shenzhen and comprehensive private tourism real estate projects in Huizhou, China. Design philosophy: Question, change, interpret.

Philip Mitchell

Philip Mitchell Design Inc. Toronto, Ontario & Chester, Nova Scotia Canada. A boutique firm specialising in residential design. Recent work includes a contemporary townhouse in Toronto, a beachfront compound in the Caribbean and a traditional apartment in New York. Current projects include a new home at the Ritz Carlton development in Naples Florida, a contemporary apartment in Toronto and a historic restoration in St. John's Newfoundland. Design philosophy: creating comfortable and beautiful living spaces that complement the personality and lifestyle of each client.

John Robert Wiltgen

Company: John Robert Wiltgen Design and JRWD Nigeria Limited. Architectural, landscape and interior design with offices in Chicago, Los Angeles and Lagos. Recent projects include a 5,000 square foot sports wing, an 86th floor penthouse at Trump International Hotel and a 7,000 sq ft penthouse for a Hollywood A List actor. Current projects include a 50,000 sq ft estate in Lagos, Nigeria, a 6,700 sq ft penthouse residence for a television talk show host and a hotel/condominium/boutique mall development in Lagos.

Collection Privée

185

Designers: Nicolette Schouten and Marianne Pellerin. Company: Collection Privée, Cannes, France. A team of 10 interior designers who specialise in the renovation and full decoration of apartments and houses in France as well as worldwide. Every project is unique and eclectic, with specially selected decoration and art pieces. Recent work includes a 1,000 sq m sea front house in St Barths and an apartment on the croisette in Cannes with the first indoor pool to be installed. Other projects have been completed in Ibiza, Cannes, St Tropez, Moscow, St Petersburg, Denmark, London, Gstaad and the Netherlands. The company's strength is delivering projects where the customer just needs to bring his clothes and enjoys the rest.

Taylor Howes Design

Designers: Karen Howes & Sandra Dreschler. Company: Taylor Howes Design, London. Founded by Karen Howes, the practice deliver high end projects worldwide. Recent work includes an 18,000 sq ft house in Connecticut, a holiday home in Mougins, France and an exclusive apartment in Lowndes Square. They are currently working on a church conversion in Knightsbridge, the largest house in London and a house in Mayfair. Design philosophy: approachable, stylish and savvy.

Fantastic Design Works

Designers: Katsunori Suzuki & Eiichi Maruyama. Company: Fantastic Design Works Co, Tokyo, Japan. Work is wide ranging, with a focus on vivid spaces that surprise. Current projects include the night club Le club de Tokyo, an Alice in Wonderland themed restaurant in Tokyo and a restaurant Hishimeki in Kyoto. Their design philosophy: to create glamour and luxury with comfort.

Claudia Pelizzari

Claudia Pelizzari Interior Design, Brescia, Italy. Composed of architects, interior designers and lighting designers who specialise in tailor made residences and hotels worldwide. Current projects include buildings in Rome, Milan, Palermo, Brescia and Naples as well as a villa on the Côte d'Azur and a chalet in Gstaad. Recent work includes a boutique hotel in Paris, a Villa in Positano and another on Lake Garda. Design philosophy: to preserve the old and equip the new.

201

David Scott

Company: David Scott Interiors, New York, U.S.A. Recent work includes a contemporary sea front home in Southampton, New York, a Georgian estate in Greenwich, Connecticut and a high-rise hotel condominium residence in Miami Beach, Florida. Current projects include a hillside resort home in Paradise Valley, Arizona, a loft in TriBeCa, New York and a complete renovation of a Villanova, Pennsylvania estate for existing clients. David's design philosophy is to create visually stimulating, yet highly functional interiors that gracefully meld practical architecture with unique design.

Design Apartment

Designer: Tang Chung Han (TT) Company: DA Design Apartment Co. Ltd, Taipei City, Taiwan. Established in 2003, consisting of a team of 20 designers who specialise in award winning, tailor made residential and commercial design throughout Taiwan. Design philosophy: to pursue a new expression of contemporary living for the future.

Jan Reuter

Jan Reuter Einrichtungen, München, Germany. Bespoke residential and commercial interiors. Recent projects include the sympathetic renovation of a 1920's country house in the Valley of Tegernsee, Germany, the restoration of the Art Nouveau style Goethe-Institut on the banks of the river Vltava, Prague and Galerie Nusser Baumgart, Munich. Current work includes a private home in the Bavarian Alps, the headquarters of 'Bayerische Staatsforsten', Regensburg and a private house in London.

Forward Architecture & Interior Design

Designers: Marco Teixeira & Alexandre Lima. Company: Forward Architecture & Interior Design, Lisbon, Portugal. An international brand of architecture and interior design specialising in timeless, contemporary work with subtle luxury. Recent projects include an apartment and a private house in Lisbon, a penthouse in São Paulo and private houses in Lisbon and Angola. Design philosophy: 'Comfort does not lie in trivial affluence but in delicacy and uniqueness. Look back to go Forward.'

Andgela Konevnina

Designer: Andgela Konevnina, Moscow, Russia. Specialising in private residential projects in Russia, the UK and Europe. Recent work includes a country house in the Moscow suburbs, a classical 19th century country manor and an apartment in Moscow. Current projects include a traditional Italian style villa in Forte, the reconstruction of an 18th century villa with vineyard in the French mountains and the renovation of a house in Holland Park, London.

Nicky Dobree

Nicky Dobree Interior Design Ltd, UK. Specialising in luxury ski chalets and high end residential projects internationally. Recent work includes a villa in Italy, a chalet in Switzerland and the renovation of a 19th century London mews. Current projects include chalets in St Moritz, Gstaad, Verbier, Klosters and Crans Montana. Design Philosophy: 'create a home that is above all comfortable, harmonious and timeless.'

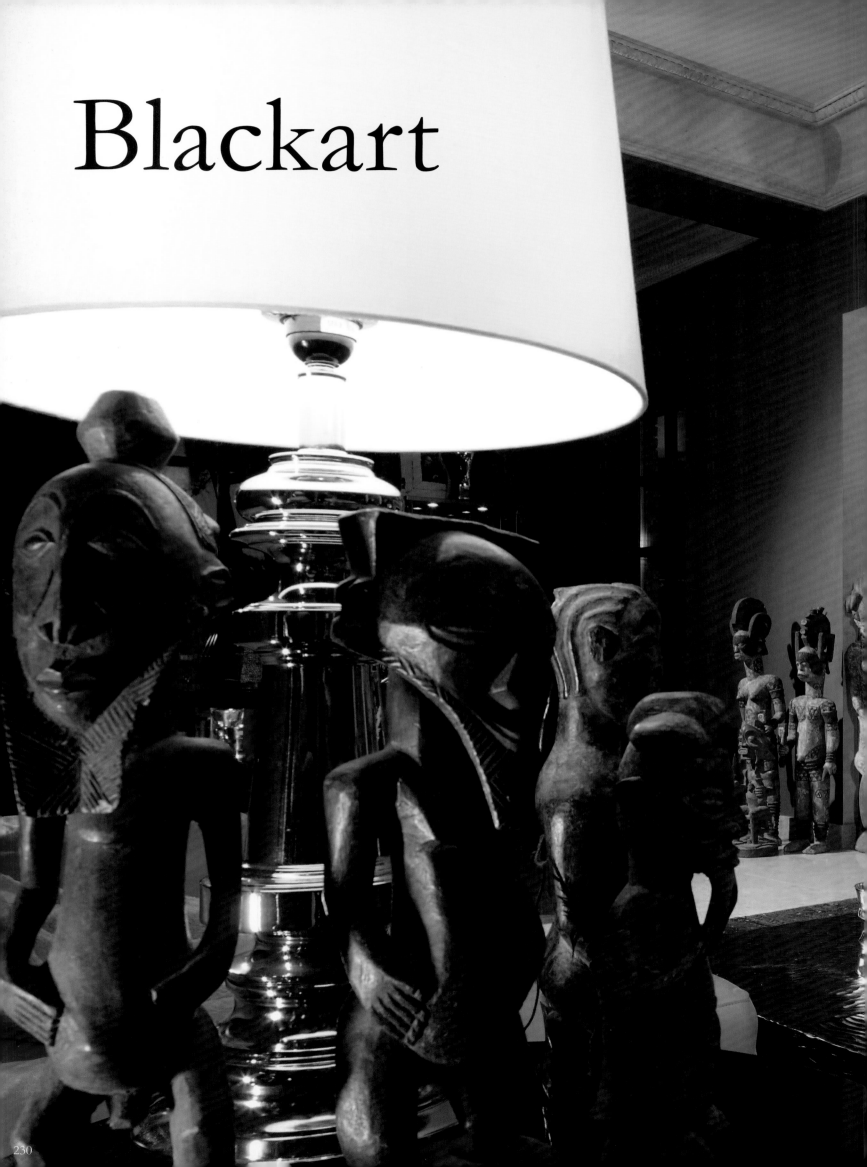

Blackart

Designer: Oksana Zakharova, Moscow, Russia. Company: Blackart. A graduate of the Design School 'Details' in Moscow, Oksana also trained at KLC, London. Work is predominantly private, specialising in country houses and apartments. Recent projects include a studio in Moscow, a house in Spain, a cottage by the lake in the Tver region and a café and karaoke club in Moscow.

Designers: Jianguo Liang, Wenqi Cai, Yiqun Wu, Junye Song, Zhenhua Luo, Chunkai Nie, Eryong Wang. Company: Newsdays, Beijing. Committed to providing their customers with solutions beyond their expectations. Projects are varied including high-end hotels, clubhouses, cultural and commercial spaces. Current work includes Beijing Yufu 77th, Capital and Investment Co., Ltd; Chongqing Liangjiang clubhouse & Nanjing Old house clubhouse, Catic Real Estate Co., Ltd.

Newsdays

Melanie Rademacher

Designer: Melanie Rademacher. Company: Mena Interiors Ltd, London. Established in May 2003, Mena Interiors has rapidly grown into a dynamic and successful interior design studio. Work is predominantly high end residential, both new build and refurbishment, tailor-made for each client. Recent projects include a converted school house in London, transforming the space into residential use whilst incorporating the owner's extensive art collection. Melanie's design aesthetic is inspired by her client's lifestyle and taste, its success is measured by how it complements the client's everyday life.

João Mansur

João Mansur Architecture & Design, São Paulo SP Brazil. Specialising in residential and commercial projects. Current work includes the restoration of a XIX century farm in upstate São Paulo, a boutique hotel in Porto, Portugal and a beach front house in Trancoso, Bahia, Brazil. João's philosophy is to balance good elements of the past with cosmopolitan and timeless accents. His motto is symmetry, proportion and quality.

Designer: Simon Mcilwraith. Company: Collective Design, Newcastle Upon Tyne, UK. Established by Simon, Collective specialise in luxury commercial design including bars, restaurants, hotels and retail environments. Current projects include a bar in London, a golf resort, sports village in the North of England and an office complex in Newcastle. Recent work includes a private residence in Northern India, a boutique hotel in Scotland and a quirky tea-house in Northumberland. Simons design philosophy is to break the mould and redefine expectation. With a firm belief that beauty and purity are timeless and that trends should be created and not followed.

Marina Filippova

Marina Filippova Designs, Moscow, Russia. Predominantly residential work internationally. Marina collaborates with fine artists worldwide to strike a balance between classic and contemporary design. Recently completed projects are in London, Switzerland and Moscow. Current work includes a mansion in London, a chalet in the Swiss Alps and apartments in Moscow and St Petersburg. Design philosophy: carefully edited interiors that combine elegance with vintage and antique design.

Stefano Dorata

Designer: Stefano Dorata. Company: Stefano Dorata Architetto, Rome, Italy. Specialising in private houses, villas and hotels predominantly in Italy but also in Europe, North and South America and the Middle East. Recent work includes a villa in Florence, an apartment in Piazza di Spagna, Rome and a flat in New York. Current projects include a house in London, a beachfront house in Sabaudia and a boutique hotel in Tel Aviv. Stefano's design philosophy is to look for classicism in contemporary life.

Hill House

Designers: Jenny Weiss & Helen Bygraves. Company: Hill House Interiors. Multi award winning practice specialising in the top end of the residential market. Recent projects include a large country estate, an Art Deco inspired villa in Cap d'Antibes and the refurbishment of a five storey house in Mayfair. Current work includes a super yacht, a luxury boutique in Chelsea and a 30,000 sq ft residence on the exclusive Wentworth estate. Design Philosophy: 'You only get one chance to make a first impression'.

Christian's & Hennie

Designer: Helene Hennie. Company: Christian's & Hennie, Oslo, Norway. The firm consists of seven interior architects and designers and two managers. Christian's & Hennie were founded in March 2007 although Helene Hennie has been in the industry for over 25 years and is highly sought after. She and her team received 'The Andrew Martin International Interior Designer of the Year Award' in 2007. Helene has also designed a furniture collection for the furniture manufacturer Slettvoll. Work is high end, with exclusive clients throughout Norway as well as internationally. Projects include private homes, summer cabins and winter lodges, to more industrial work such as restaurants, hotels and offices. Current projects include a mansion in the USA, a small, historical private house in Oslo and a farmhouse in Italy.

275

Designer: Brian J. McCarthy. Company: Brian J. McCarthy Inc., New York, USA. Brian joined Parish-Hadley Associates in 1983 as assistant to the venerable Albert Hadley where he eventually became a full partner before starting his own firm in 1991. Recent residential projects include residences in Palm Beach, Bel Air and on Long Island, apartments in Manhattan plus the private offices for the CEO of a major NYC based private equity firm. He was also hired to redecorate the private quarters and to refurbish the official State Rooms in Winfield House, the United States ambassadorial residence, set in

Regents Park, London. Current projects include a 70,000 sq ft ski chalet in Gstaad, a family compound on the coast of Maine, a Chicago penthouse overlooking Lake Michigan as well as a Park Avenue triplex and a weekend residence in Water Mill, NY. Brian's style is rooted in classical European design while incorporating a modern American sensibility.

Debra Cronin

Debra Cronin Design, NSW, Australia. A boutique practice appealing to those seeking a quirky and idiosyncratic space, where nothing is off limits. Appearing in countless interior magazines and books around the world, recent work has included an award winning restaurant in Potts Point and a fashion designer's home in Bellevue Hill. Current projects include a landmark property perched on a Central Coast cliff top and a city warehouse conversion. Debra's distinctive trademark marries her madcap aesthetic with an impressive knack of hunting down the sublime and the obscure, for both contemporary and historic interiors.

Nerija R. Sabaliauskiene

Designer: Nerija R. Sabaliauskiene. Company: Nerija Interior Design, Lithuania. Specialising in luxury residential and commercial interiors around the world. Current projects include seaside villas in Tuscany and Neringa as well as several large family houses in the most exclusive areas in Vilnius, Lithuania. Recent work includes the interior of a private jet, a contemporary house for a football star and a family house in London. Design philosophy: combining old and new elements to create intimate design.

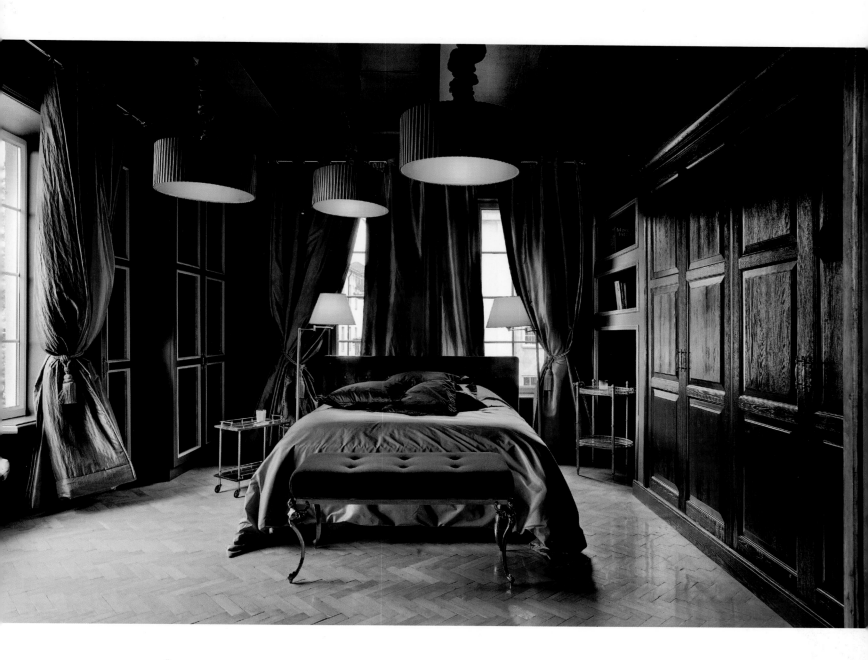

François Champsaur

293

Designer: François Champsaur. Company: Agence François Champsaur, Paris. A French interior designer who runs a medium sized practice specialising in high end hotels and residential projects in France as well as in the United States, China and Mauritius. François Champsaur designs not only the interiors but also the furniture for each project. Current work includes the complete renovation of the 5 star Hôtel Royal Evian, Hôtel Regina Eiffel, Paris and a 440 sq m flat in Paris. In 2011 they completed the transformation of a 24-room boutique hotel in the 8th district of Paris, the new Club Med Sandpiper resort in Florida and signed the opening of the Yearlings equestrian auction hall in Deauville. Design philosophy: to blend contemporary furniture and fabrics, marrying sophistication, comfort and poetry.

Jeffreys Interiors

Designers: Alison Vance and Jeff Laing. Company: Jeffreys Interiors, Edinburgh, UK. Recent work includes a Scottish Castle, a small Highland Hotel and a beach house in North Berwick. Current projects include a boutique hotel in Edinburgh, a Georgian property in Bristol and an underground car park conversion into a VIP Bar for the Edinburgh Festival.

303

Kira Chuveleva

Company: Chuveleva Interior Design, Moscow, Russia. Specialising in contemporary residential and small scale commercial interior design. Kira's inspiration is Bauhaus design principles, mid 20th century Scandinavian design, Russian constructivist architecture and modern art. Recent projects include several apartments in downtown Moscow, an apartment in a 19th century house in the city centre, a private spa and an office in the Moscow suburbs. Design philosophy: combine new and old and above all reflect the client's personality.

307

Alla Shumeyko

Designer: Alla Shumeyko. Company: Alla Decor Studio, Moscow. Specialising in the decoration and design of private and public interiors including furniture and accessories. Recent projects include a private apartment and residence. Current work includes a country house and apartments. Design philosophy: harmony and comfort.

Linda Steen

Designer: Linda Steen. Company: AS Scenario Interior Architects MNIL, Oslo, Norway. Scenario's mission is to design the best of tomorrow's environments. Specialising in various ongoing high profile projects. Recent work includes Deichmanske, Oslo's new Public Library, Rica Hotels and Siemens and Aker Solutions offices. Their design philosophy: functionality with comfort.

Kris Lin

Designer: Kris Lin. Company: KLID (Kris Lin Interior Design) Shanghai, China. Recent work includes various Sales Centres in Wenzhou and Rizhao, China, and in Shanghai, Club houses in Shenzhen, and Guangzhou and villas in Fuzhou.

Etcetera Living

Designer: Lesley Zaal. Company: Etcetera Living, Dubai. Established in 2003 to service the needs of a rapidly growing population in Dubai, the company has grown with the city and has completed many high end residential and commercial projects. Recent work includes a luxury beachfront residence in Abu Dhabi, a villa on The Palm and several houses in Al Barari. Current projects include 'The Farm', an all day dining luxury restaurant in Dubai and a private house and a boutique hotel in Abu Dhabi.

Lígia Casanova

Designer: Lígia Casanova. Company: Atelier Lígia Casanova, Portugal. Founded in 1994 with the aim of creating projects that follow the company motto 'to make room for happiness', whether commercial, residential or public space. Recent projects include luxury apartments in Lisboa, Cascais and Oporto.

Jimmie

Martin & McCoy

Designers: Jimmie Karlsson, Martin Nihlmar and Sally Anne McCoy. Company Jimmie Martin and McCoy, London. A young design practice working in residential and commercial design including nightclubs, boutique hotels and offices. Recent projects include offices in Notting Hill, a loft apartment in Kensington and a nightclub in London's West End. Current work includes a warehouse apartment in Soho, a country boutique guest house and a penthouse in London's Kings Cross. Design philosophy: 'luxury without rules.'

If you're easily offended — Now would be a good time to fuck off.

Michael Attenborough

Designer: Michael Attenborough. Company: Radisson Blu Edwardian. A collection of 13 hotels, each with its own unique character and an uncompromising blend of design, comfort and technology. Situated in the heart of the Mayfair village, the 406-bedroomed May Fair Hotel has been a by-word for elegance and style since it was opened by King George V. It is the official hotel of London Fashion Week and the BFI London Film Festival. Distinguished by its 13 Signature Suites, The May Fair Hotel has also been recognised for having "the Best Hotel Bar in London" (Evening Standard) and awarded Cool Brands status in 2010 and 2011. Michael Attenborough's design philosophy is bold, contemporary and without compromise.

aa studio

Designers: Roger Pop & Alex Adam. Company: aa studio, Bucharest, Romania. Established in 2004 by Alex Adam and Roger Pop, the studio has been awarded both in national and international architectural competitions and has been responsible for the design of over 35 projects in Romania. Specialising in residential and commercial interior design their work has also been published in numerous domestic as well as international magazines. Design philosophy: adaptability.

345

Rosa May Sampaio

Designer: Rosa May Sampaio. Company: Rosa May Decoração de Interiores, São Paulo, Brazil. Predominantly residential work including offices, clubs, shops and restaurants. Current work includes a penthouse in Ipanema, Rio de Janeiro, another in São Paulo plus houses, farms, apartments and a restaurant. Rosa May's design philosophy is above all to provide comfort and harmony.

Abraxas

Designer: Christian Baumann. Company: Abraxas Interieur GmbH, Zurich. Since 1995 Christian has specialised in high end, bespoke work internationally. Recent projects include a mountain apartment in St. Moritz, the concept and interior design for the top floor of a house on Zurich's lake and the interior design for a modern sea front apartment. Current projects include the concept and interior design of a members club in the Swiss mountains, the interior design for an original Bauhaus residence in downtown Zurich and the full concept and interior design for a new build at Zurich's gold coast. Design philosophy: 'to make the ordinary extraordinary.'

Company: Steve Leung Designers, Hong Kong. Leading International architect, interior designer and product designer, born in Hong Kong in 1957. Steve set up his architectural and urban planning consultancy in 1987. Projects are varied, including hotels, restaurants, show flats, club houses and retail. His style is contemporary minimalism with the adoption of Asian culture and arts. Recent work includes Sing Yin at W Hong Kong, The Hampton, 39 Conduit Road and Mango Tree in Hong Kong. Current projects include a hotel in London, a residential development project in Singapore and a resort hotel in Sanya, China. The company has been credited with over 80 design and corporate awards in the Asia Pacific region and worldwide. Steve's philosophy: 'Enjoy Life. Enjoy Design.'

Intarya

Designer: Daniel Kostiuc and team. Company: Intarya, London. Intarya create elegant, functional and comfortable interiors worldwide. Recent projects include an apartment in Bangkok, a country house in Dublin and two apartments overlooking Hyde Park, London. Current work includes a palace in Saudi Arabia, a Grade II listed country house in Devon and a pied-à-terre in Belgravia. Intarya's designs provide quality, timelessness and understated luxury.

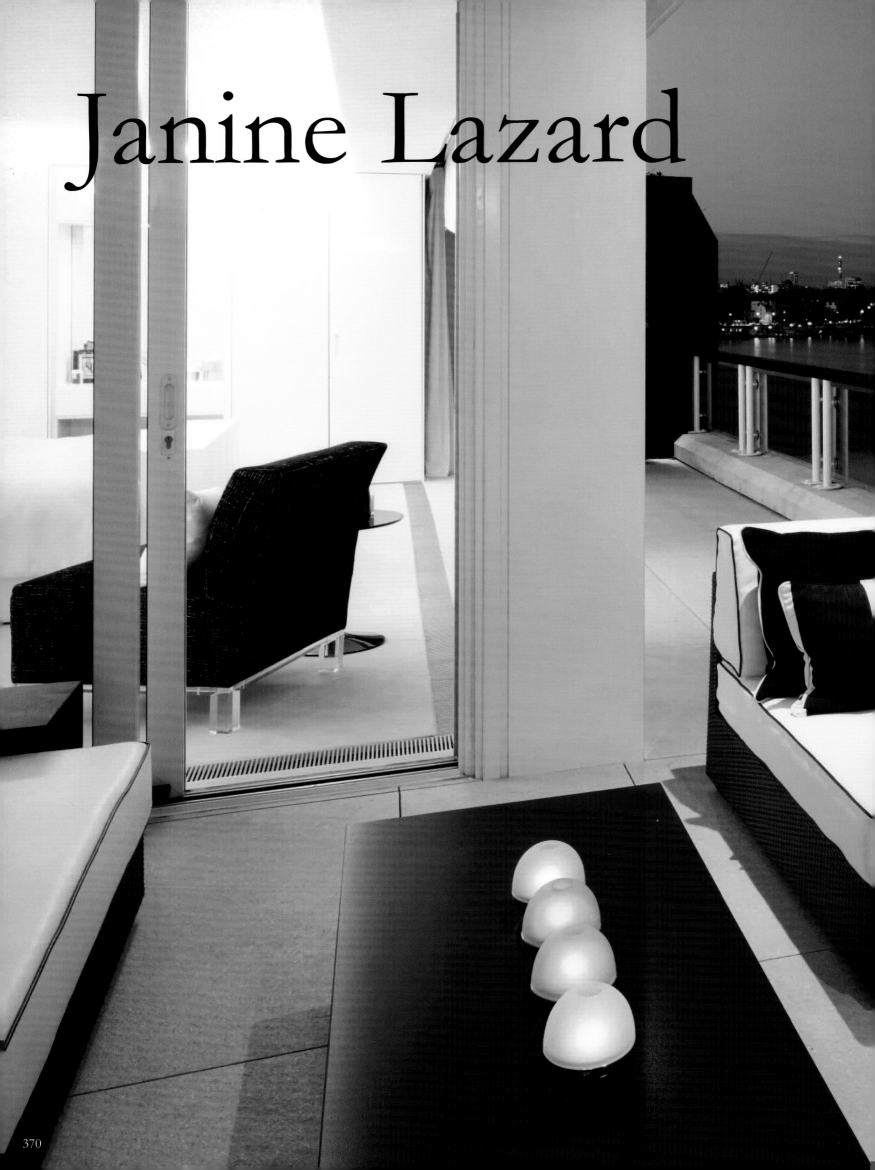

Janine Lazard

Designer: Janine Lazard. Company: Janine Lazard Interiors. Specialising in high end residential projects globally with offices in Johannesburg, Cape Town and London. Current projects include an apartment in Regents Park, London, a luxury home in Vaucluse, Sydney, Australia and a penthouse in Michelangelo Towers, Sandton, Johannesburg. Recent work includes an apartment in Los Angeles, a penthouse in Chelsea, a luxury home in Manchester for an England footballer and a Clifton Villa in Cape Town owned by Kingfisher boss Dr Vijay Mallya.

Ivan C. Design

Designer: Ivan Cheng. Company: Ivan C. Design Limited, Hong Kong. Specialising in hotels, villas, sales offices, show flats and clubhouses. Recent work includes Sky City Marriott Hotel, Hong Kong International Airport, Hilton Hangzhou Qiandao Lake Resort, Bin Sheng Xiang-Lake Villas and clubhouse, Hangzhou. Current projects include Li Shui Xin Hu show flats and clubhouse, a Hong Kong Private Art Museum and Maestro Liu & Sun - The Club Extraordinaire.

Joanna Berryman

Designer: Joanna Berryman. Company: Matrushka, London, UK. Predominantly luxury, residential property in London, with expansion this year into the commercial arena. Recent work includes the refurbishment of a stucco fronted villa and a Gothic Lodge plus an Edwardian conversion. Current projects include an office space, a Georgian townhouse and a large Victorian villa. Matrushka incorporate instinctive and witty design with the traditional and modern.

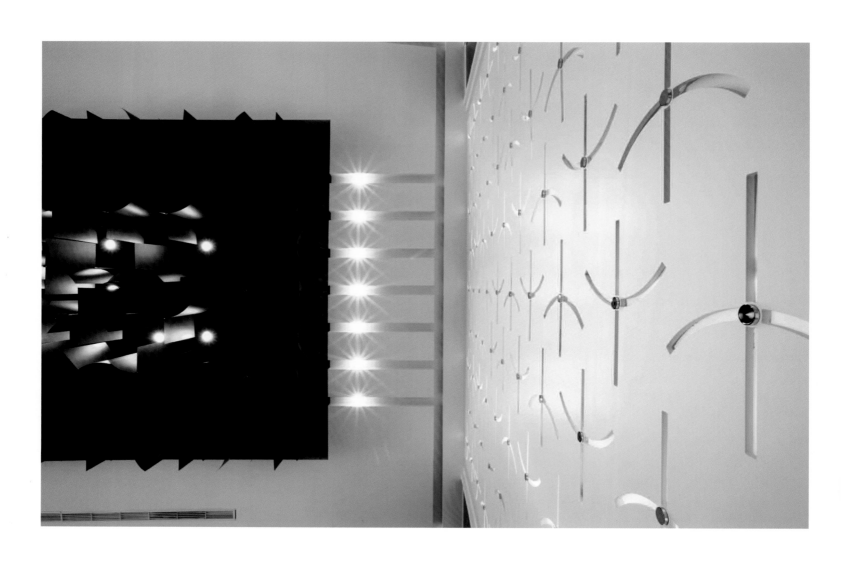

Chang Ching-Ping

Designer: Chang Ching-Ping. Company: Tien-Fun Interior Planning, Taiwan, China. Established in 1988 at Tienmu, Taiwan and now based in Taichung. Tien Fun specialise in high end residential and show flat design throughout Asia. Their team consists of more than 20 experienced designers working in various differing styles to suit each client brief. Design philosophy: high quality design and craftsmanship.

Olga Stupenko

Designer: Olga Stupenko. Company: OS Empire of Design, Moscow, Russia. Private and commercial high end interiors, predominantly in Russia. Recent work includes the Hotel La Farandole on the French coast and apartments in Moscow. Current work includes four houses near Moscow. Olga's vision is to create honest, timeless interiors with a sense of humour.

Anemone Wille Våge

Anemone Wille Våge Interior Design, Oslo, Norway. Work is residential and commercial including hotels, restaurants, offices as well as private residences and chalets. Current projects include The Thief Hotel in Oslo, Hotel Union Øye in Ålesund and Highland Lodge at Geilo. Previous projects include Hotel Post in Sweden, a large winter chalet in the mountains of Norway and a summer residence by the south coast of Norway. Design philosophy: tailor made design in harmony with its architecture and surroundings.

Joseph Sy

Joseph Sy & Associates. Specialising in a wide variety of projects including restaurants, bars, retail, corporate and residential plus religious spaces. Current work includes restaurants in Lanzhou and Guangzhou and serviced apartments in Hong Kong. Design philosophy: function prior to aesthetics.

Kari Arendsen

Principal Interior Designer: Kari Arendsen & Associate Interior Designer: Jenn Bibay. Company: Intimate Living Interiors, Solana Beach, CA, USA. Recent work includes a 4,000 sq ft seaside villa in La Jolla, California, a 3,800 sq ft colonial bachelor retreat for an investment banker in La Jolla, plus a 4,500 sq ft ranch and coastal dwelling for a professional athlete and young family in Rancho Santa Fe, California. Design philosophy: blend reclaimed with refined and the unexpected with the essential.

Spinoff

Designers: Ichiro Shiomi and Etsuko Yamamoto. Company: Spinoff Co., Ltd, Tokyo, Japan. Projects are wide ranging including commercial, retail and residential design. Recent work includes a Japanese restaurant in Skytree Tower Tokyo, a chocolate retail shop in Tokyo and a Japanese restaurant in Seoul, Korea. Current projects include a chain of coffee shops in Korea, a large shopping centre space design in Seoul and a Spanish restaurant in Tokyo. Design philosophy: creating atmospheric spaces that capture people's hearts.

John Rufenacht

Designer: John G. Rufenacht. John Rufenacht Associates, Inc. Kansas City, Missouri, USA. A small company with major residential interiors in Pebble Beach and Long Beach, California to Aspen and Denver, Colorado to Houston and Dallas, Texas. Current projects include Palm Desert, California. They believe 'home is ones only true sanctuary and that it should nourish the soul.'

Thomas Dariel

Designer: Thomas Dariel. Company: Dariel Studio, Shanghai. An international studio founded by Thomas Dariel. Recent projects include Blossom Hill Heritage, a boutique hotel in Zhouzhuang, Dunmai office in Shanghai and Kartel Wine Lounge in Shanghai. Current work includes the renovation of Mix Club in Beijing, a three-floor office building in Shanghai and a 1,500 sq m apartment in Beijing. Design philosophy: to give a life to space, with boldness, elegance and quality.

Robert Mills

Designer: Robert Mills. Company: Robert Mills Architects, Melbourne, Australia. Specialising in contemporary residential architecture and interior design. Recent projects include an award winning villa, the heritage restoration of a mansion estate and a beach house offering all the sophistication and convenience of modern life but cloaked in casual clothes. Current projects include a cliff top beach house inspired by the Capri coast, a contemporary urban residence overlooking Melbourne's city skyline and a cutting edge family residence in Brighton, Victoria. Rob believes contemporary residential design should be sophisticated and relaxed, creating exceptional living spaces.

Fabio Galeazzo

Designer: Fabio Galeazzo. Company: Galeazzo Design, São Paulo, Brazil. A multidisciplinary team specialising in architecture, interiors and product design. Recent work includes the complete reconfiguration of a farmhouse plus private residences. Current projects include a restaurant, a contemporary penthouse and a chapel in bamboo. His design philosophy: 'make the world look better.'

TF Designs Global

Designers: Miss Tang Chih-Li, Mr. Yeh Chun-Erh, Mr. Law Louis. Company: TFDG, Taiwan. Founded in 2010, specialising in showflats, club houses, landscape design, town planning, residential and furniture design and office design in Taiwan, China and Singapore. Recent projects include a residence in Singapore, a sales office at Da Lian Geng Hai, Haleakala Hawaii Restaurant and the Club House of Mansion De Crillon. Design philosophy: quality, craftsmanship, professionalism.

Sonia Korn

Sonia Korn Developing & Design Ltd, New York & London. Projects are predominantly new build with a focus on finding the right space in the desired location anywhere in the world, then creating the client's dream. Recent work includes a country house and a townhouse in London and an apartment in Manhattan. Current projects include a house and a spa in Manhattan and a night club in London. Sonia's preference is for eclectic spaces, bringing outside to inside.

Ryu Kosaka

Designer: Ryu Kosaka. Company: A.N.D., Tokyo, Japan. Specialising in high end commercial, residential and product design predominantly in Japan. Recent work includes a hotel lounge in Macau, a Japanese restaurant at the Palace Hotel Tokyo and 6th by Oriental Hotel in Yurakucho Tokyo. Motto: 'everlasting design'.

Kathryn M. Ireland

Company: Kathryn M. Ireland Textiles & Design, Los Angeles, USA. British-born Kathryn moved to Los Angeles in 1986 and began her business in the early 90's. Her first fabric collection was launched in 1997 and today her collections are distributed through her showrooms in Los Angeles and London and around the world. Originally Kathryn was an actress, clothing designer and filmmaker. She has also written four books: Timeless Interiors, Classic Country, Creating a Home and more recently Summers in France. Clients are predominantly celebrity including Steve Martin, Julia Louis-Dreyfus and Arianna Huffington. Kathryn's style is comfortable, colourful, bohemian and unorthodox.

455

Staffan Tollgård

Designer: Staffan Tollgård. Company: Staffan Tollgård, London, UK. Listed on the House & Garden's 100 Leading Designers directory. Brought up in Stockholm, Staffan worked as a film producer before enrolling on the Postgraduate Diploma in Architectural Interior Design. He worked for designer Rabih Hage before starting his own company in 2005. Based in Notting Hill, the group work on a wide range of projects, offering a bespoke interior design service to an international clientele. In February 2010 Staffan also launched a design studio in the Algarve. Current projects include a Belgravia townhouse, a Georgian manor house on the outskirts of Windsor's Great Park and a family home in Lagos, Nigeria. Recent work includes two contemporary villas in the Algarve, an Arts and Crafts inspired mansion in Surrey and several private homes in central London. Staffan's style is borne out of Asian and Scandinavian functionalism, combined with a strong appreciation of furniture as sculptural art.

Peter Phan

Company: Peter Phan Design Consultancy, London, UK. Established in 2008, with an office in Marylebone. Recent work includes two private houses in Regent's Park, a country house in Oxfordshire, a Chateau in France, a golf course clubhouse in France and three private adjoining Grade 1 listed houses in Regent's Park. Design philosophy: modern classic.

467

Angelos Angelopoulos

Designer: Angelos Angelopoulos, Athens, Greece. In 1990 Angelos designed his first hotel, Andromeda Athens. Since then he has designed the interiors of more than 40 predominantly boutique hotels. Apart from Greece, projects are in Cyprus, Istanbul, New York and lately in Florida and New Jersey. They include residences, apartments, restaurants, clubs, hotels and showrooms. Recent work includes full conceptual designs for buildings as well as hotels and private residences in Greece and Cyprus and restaurants in the USA. Design philosophy: 'social dynamics, observation, interpretation. The most significant part of Angelos's work is psychology; the interior designer is invited to cover four basic human needs: change, distinctiveness, communication, safety.'

471

Greg Natale

Company: Greg Natale Design, Sydney, Australia. Having studied both interior design and architecture, Greg's holistic approach and signature style has seen his business thrive for over a decade. Recent projects include a waterfront apartment on Sydney Harbour, a country house in Sutton Forest and the remodelling of a Modernist house in Brighton Melbourne. Current work includes the construction of a new glass and concrete tri-level home on Sydney Harbour, a riverside home in Sydney's Northern beaches and a golf course property on the Royal Pines Resort, Gold Coast. Design philosophy: be bold.

Yvonne O'Brien

Designer: Yvonne O'Brien. Company: The Private House Company, South Africa. Work is residential and commercial including more recently a private beach house in Machangulo, a boat house on the Zambezi River and two private residential homes on a Golf Estate. Other projects include Londolozi in the Sabi Sands Game Reserve, a lodge in the Namib Desert and a residential home on a Polo Estate. Design philosophy: understated, elegant, comfortable.

PTang Studio

Designers: Philip Tang & Brian Ip. Company: PTang Studio Ltd, Hong Kong. Specialising in residential and commercial design. Recent projects include a residential development in Dubai Pearl, Dubai, a luxury residential development in Tai Po, Hong Kong and a high-rise residential building in Dalian, China. 'PTang Studio Ltd is developing a fresh and unique style that transcends existing boundaries and widens the horizon of design in the visionary future.'

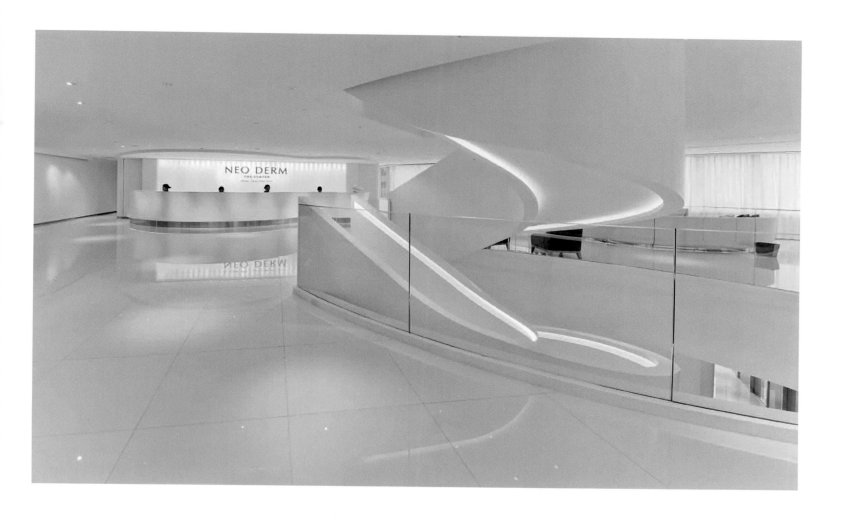

SA Aranha & Vasconcelos

Designers: Carmo Aranha and Rosário Tello. Company: SA, Aranha & Vasconcelos, Lisbon, Portugal. Established in 1986, SA & V is an architecture and interior design studio based in Lisbon. Recent projects include a 43 metre yacht, Baglietto, presented at last September's Monaco Yacht Show and 2 apartments in Lisbon. Current work includes the interior design for a complex of houses in a large rural property in Alentejo, Portugal, a shoe shop in Zurich and a whole building for a single family in Lisbon. Design philosophy: dream to achieve different results in every project.

493

Christopher Dezille

Designer: Chris Dezille. Company: Honky, London. An award winning practice which provides a full service, specialising in interior architecture and design for private clients, developers and hospitality sectors. Recent work includes numerous properties for private clients and developers throughout London, a refurbishment in Hampstead and a family home in Surrey. Current projects include apartments in St Giles & Trafalgar Square, a residential building in Knightsbridge and a boutique hotel in Montenegro.

Slettvoll

Designer: Liv Norun. Company: Slettvoll Mobler AS, Norway. Comprising 18 shops with a design service in Norway and Sweden, Slettvoll also carry a line of furniture, carpets, lighting and a broad range of interior items. Their aim is to create homes that their clients long to go back to, relaxing and comfortable spaces of quality, timelessness and durability.

504

Zeynep Fadillioglu

Company: Zeynep Fadillioglu Design, Istanbul, Turkey. Founded by Zeynep in 1995, ZFD specialise in a broad range of projects including private houses, mosques, yachts and commercial commissions all around the world. Her work is renowned for its colour, texture, textiles and architectural features, her style combines traditional with contemporary for universal appeal.

4 Geoffrey Bradfield
Geoffrey Bradfield Inc
116 East 61st Street
New York NY 10065
Tel: +1 212 758 1773
Fax: +1 212 688 1571
info@geoffreybradfield.com
www.geoffreybradfield.com

12 Abu Jani & Sandeep Khosla
Abu Jani Sandeep Khosla
A 23, Ghanshyam Industrial Estate
Andheri West
Mumbai 58, India
Tel: +91 22 26732213
info@abusandeep.com
www.abusandeep.com

22 Jorge Cañete
Jorge Canete Interior Design Philosophy
Château d'Hauteville - Aile Est
CH 1806 St Legier (VD)
Switzerland
Tel: +41 78 710 25 34
Fax: +41 21 944 37 57
info@jorgecanete.com
www.jorgecanete.com

30 Juan Montoya
Juan Montoya Design Corp
330 East 59th Street
New York NY 10022, U.S.A.
Tel: +1 212 421 2400
Fax: +1 212 421 6240
info@juanmontoyadesign.com
www.juanmontoyadesign.com

38 Meryl Hare
Hare + Klein
Level 1, 91 Bourke Street
Woolloomooloo NSW 2011
Australia
Tel: +61 2 9368 1234
Fax: +61 2 9368 1020
info@hareklein.com.au
www.hareklein.com.au

44 David Rockwell
Rockwell Group
5 Union Square West, 8th Floor
New York, NY
10003 USA
Tel: +1 212 463 0334
Fax: +1 212 463 0335
pr@rockwellgroup.com
www.rockwellgroup.com

52 Joseph Karam
Galerie Joseph Karam
61 Avenue Raymond Poincaré
75116 Paris, France
Tel: +33 (0)1 44 05 06 06
Fax: +33 (0)1 44 05 09 20
jkaram@josephkaram.com
www.josephkaram.com

58 Heidrun Diekmann
Heidrun Diekmann Interior Lifestyles CC
PO Box 24217
162 Mugabe Ave
Windhoek, Namibia
Tel: +264 61 240 607
Fax: +264 61 240 609
hdlifestyles@diekmannassociates.com

66 Ajax Law Ling Kit & Virginia Lung
One Plus Partnership Limited
9/F New Wing, 101 King's Road
North Point, Hong Kong
Tel: +852 2591 9308
Fax: +852 2591 9362
admin@onepluspartnership.com
www.onepluspartnership.com

72 Judy Hutson
Judy Hutson
The Pig, Beaulieu Road
Brockenhurst, New Forest
Hampshire SO42 7QL
Tel: +44 (0)1590 662 354
info@thepighotel.com

78 Jay Jeffers
Jeffers Design Group
1035 Post St. San Francisco
California 94109, USA
Tel: +1 415 921 8880
Fax: +1 415 921 8881
info@jeffersdesigngroup.com
www.jeffersdesigngroup.com

86 Silvio Rech & Lesley Carstens
Silvio Rech & Lesley Carstens Architecture
& Interior Design
32B Pallinghurst Road
Westcliff 2103, Johannesburg
South Africa
Tel: +27 82 900 9935
adventarch@mweb.co.za

92 Sam Cardella
Cardella Design
839 Inverness Drive, Rancho Mirage
California 92270, USA
Tel: +1 760 324 7688
Fax: +1 760 324 7728
cardelladesign@msn.com
www.cardelladesign.com

100 Shi Xudong
Xuridongsheng Decoration Organisation
A5 Design Centre
4th Xin Gui Residence no 316
Hudong Road
Fujian, China
Tel: +86 (591) 876 722 23
Fax: +86 (591) 876 722 73
www.xrds888.com

106 Kirk Nix
KNA Design
8255 Beverly Blvd. # 225
Los Angeles
California 90048, USA
Tel: +1 323 944 0100 ext: 17
Fax: +1 323 944 0105
julia@knadesign.com
www.knadesign.com

114 Graça Viterbo & Gracinha Viterbo
Viterbo Interior Design
Avenida de Nice 68
2765-259 Estoril, Portugal
Tel: +351 214 646 240
Fax: +351 214 646 249
info@gviterbo.com
www.gviterbo.com

120 Wang Guan, Liu Jianhui, Wang Zhaobao
Matrix
20F, Building C, Fountain City Suites
Xinwan Road, Futian District
Shenzhen, China
Tel/Fax: +86 755 832 225 78
Mobile: +86 135 106 700 67
matrixdesign@live.cm
www.matrixdesign.com.cn

126 Bunny Turner & Emma Pocock
Turner Pocock Design
204 Latimer Road, Kensington
London W10 6QY
Tel: +44 (0)208 463 2390
Fax: +44 (0)208 968 7926
info@turnerpocock.co.uk
www.turnerpocock.co.uk

130 Marc Hertrich & Nicolas Adnet
Studio Marc Hertrich & Nicolas Adnet
14 rue Crespin du Gast
75011, Paris, France
Tel: +33 (0) 1 43 14 00 00
Fax: +33 (0) 1 43 88 86 01
contact@studiomhna.com
www.studiomhna.com

136 Elena Akimova
Elena Akimova
42 Bolshaya Gruzinskaya St 123056
Moscow, Russia
Tel: +7 495 967 9460
Fax: +7 499 254 8930
5261408@mail.com

142 Gluckstein Design Planning
234 Davenport Road
Toronto, Canada M5R 1J6
Tel: +416 923 6262
Fax: +416 923 6313
info@gluckseinhome.com
www.gluckseinhome.com

148 Daun Curry
Modern Declaration
37 Wall St, Suite 11B
New York NY 10005, U.S.A.
Tel: +212 480 2593
Fax: +212 937 3117
info@moderndeclaration.com
www.moderndeclaration.com

154 Feng Yu
Deve Build
W1 - 201 East Garden OCT
Nanshan District, Shenzhen, China
Tel: +86 755 266 038 06
Fax: +86 755 821 292 05
devebuild@yahoo.com.cn
www.devebuild.com

160 Catarina Rosas, Claudia Soares Pereira,
Catarina Soares Pereira
Casa do Passadico
Largo de S. Joao do Souto
4700 326 Braga, Portugal
Tel: +351 253 61 99 88
Fax: +351 213 110
mail@casadopassadico.com
www.casadopassadico.com.

166 Julia Buckingham Edelmann
Buckingham Interiors & Design
1820 West Grand Avenue
Chicago, Ilinois 60622, U.S.A
Tel: +1 312 243 9975
Fax: +1 312 243 9978
info@buckinghamid.com
www.buckinghamid.com

170 Bo Li
Cimax Design Engineering (Hong Kong) Ltd
7B Building 9 Haiying Chang Chen
No 2 Wenxin Road
Shen Zhen
Guang Dong China
Tel/Fax: +86 755 26448677
libodesign@126.com
www.libodesign.com

176 Philip Mitchell
Philip Mitchell Design Inc
160 Pears Avenue, Suite 302
Toronto, Canada M5R 3P8
Tel: +416 364 0414
Fax: +416 364 1206
info@philipmitchelldesign.com
www.philipmitchelldesign.com

180 John Robert Wiltgen
John Robert Wiltgen Design Inc
70 W Hubbard Suite 205
Chicago, Il. 60654 USA
Tel: +1 312 744 1151
Fax: +1 312 321 9590
info@jrwdesign.com
www.jrwdesign.com

184 Nicolette Schouten and Marianne Pellerin
Collection Privée
3 rue des Etats - Unis
06400 Cannes, France
Tel: +33 (0) 493 99 23 22
Fax: +33 (0) 493 39 99 89
cannes@collection-privee.com
www.collection-privee.com

188 Karen Howes, Sandra Drechsler
Taylor Howes
29 Fernshaw Road
London SW10 OTG
Tel: +44 (0)207 349 9017
Fax: +44 (0)207 349 9018
admin@taylorhowes.co.uk
www.taylorhowes.co.uk

192 Eiichi Maruyama
Fantastic Design Works Co Ltd
4F Sariitus Building 1-32-18 Uehara
Shibuya-ku, Tokyo
151-0064 Japan
Tel: +81 3 6659 5401
Fax: +81 3 6659 5402
tokyo@f-fantastic.com
maruyama@f-fantastic.com
www.f-fantastic.com

198 Claudia Pelizzari
Claudia Pelizzari Interior Design
Via Trieste, 41
25121, Brescia, Italy
Tel: +39 030 29 00 88
Fax: +39 33 439 54 239
info@pelizzari.com
www.pelizzari.com

202 David Scott
David Scott Interiors Ltd
1123 Broadway, 8th Floor
New York, NY 10010
Tel: +212 829 0703
Fax: +212 829 0718
info@davidscottinteriors.com
www.davidscottinteriors.com

206 Tang Chung Han
Design Apartment
No 3 Ln. 214 Rui an St
Da' an Dist, Taipei City 106
Taiwan
Tel: +886 2 2703 1222
Fax: +886 2 2703 1223
da.id@msa.hinet.net

212 Jan Reuter
Jan Reuter Einrichtungen
Therese-Studer-Straße 29 D-80797
München, Germany
Tel: +49 89 52 30 29 00
Fax: +49 89 52 30 29 29
jan.reuter@jan-reuter-einrichtungen.de
www.jan-reuter-einrichtungen.de

216 Marco Texeira & Alexandre Lima
Forward Architecture & Interior Design
Rua de Joao Penha, 14A
1250 - 131, Lisbon, Portugal
Tel: +351 213 828 400
lisbon@forward-aid.com
www.forward-aid.com

220 Andgela Konevnina
Andgela Konevnina
22 Klimashkina St
Apartment 33
Moscow, Russia
Tel: +7 916 221 9628
Tel: +3712 626 2676
andgela-boss@mail.ru

224 Nicky Dobree
Nicky Dobree Interior Design Ltd
25 Lansdowne Gardens
London SW8 2EQ
Tel/Fax: +44 (0)207 627 0469
info@nickydobree.com
www.nickydobree.com

230 Oksana Zakharova
Blackart
109145 Zhulebinsky Boulevard 5--239
Moscow
Russia
Tel: +7 495 987 42 18
oxana.06@mail.ru

236 Jianguo Liang, Wenqi Cai, Yiqun Wu, Junye Song, Zhenhua Luo, Chunkai Nie, Eryong Wang
Beijing Newsdays Architectural Design Co. Ltd.
Jia 10th, Bei San Huan Zhong Road
West City District, Beijing
P.R. China 100011
Tel: +86 10 820 869 69
Fax: +86 10 820 878 99
newsdays@newsdaysbj.com
www.newsdays.com

242 Melanie Rademacher
Mena Interiors
134 Lots Road
London SW10 ORJ
Tel: +44 (0)207 349 7115
Fax: +44 (0)207 349 7154
interiors@menainteriors.co.uk
www.menainteriors.co.uk

246 João Mansur
Joao Mansur Architecture & Design
Rua 1922b Groenlândia, Jardim América
01434 - 100, São Paulo, Brazil
Tel: +55 11 3083 1500
Fax: +55 11 3081 7732
joao@joaomansur.com
www.joaomansur.com

252 Simon Mcilwraith
Collective Design
21 Kepple Street
Dunston, Gateshead
Tyne & Wear NE11 9AR
Tel: +44 (0)191 493 2545
studio@collective-design.co.uk
www.collective-design.co.uk

256 Marina Filippova
Marina Filippova Designs
10 Vozdvishenka Str.
Moscow 125009
Russia
Tel: +7 916 146 3369
info@marina-filippova.com
www.marina-filippova.com

260 Stefano Dorata
Stefano Dorata Architect
12a/14 Via Antonio Bertoloni
Rome, 00197 Italy
Tel: +39 06 8084 747
Fax: +39 06 8077 695
studio@stefanodorata.com
www.stefanodorata.com

264 Jenny Weiss & Helen Bygraves
Hill House Interiors
32-24 Baker Street
Weybridge
Surrey KT13 8AT
Tel: +44 (0)1932 858 900
Fax: +44 (0)1932 858 997
helen@hillhouseinteriors.com
jenny@hillhouseinteriors.com
www.hillhouseinteriors.com

270 Helene Hennie
Christian's & Hennie
Skovveien 6, N - 0275 Oslo
Norway
Tel: +47 22 12 13 50
Fax: +47 22 12 13 51
info@christiansoghennie.no
www.christiansoghennie.no

276 Brian J. McCarthy
Brian J. McCarthy Inc
140 W 57th Street
Suite 5B, New York
NY 10019, USA
Tel: +1 212 308 7600
Fax: +1 212 308 4242
www.bjminc.com

282 Debra Cronin
Debra Cronin Design
56 Oxford St Woollahra
NSW 2025, Australia
Tel: +61 410 69 64 67
Tel: +61 2 8021 6440
debra@debracronindesign.com
www.debracronindesign.com

286 Nerija Sabaliauskiene
Nerija Interiors
L. Stuokos-Gucevicaius 9
Vilnius LT - 01122
Lithuania
Tel: +37 0611 331 55
Fax: +37 0523 122 44
info@nerija.co.uk
www.nerija.co.uk

292 François Champsaur
Agence Francois Champsaur
42 rue de Sévigné
75003 Paris, France
Tel: +33 (0) 1 4345 2246
Fax: +33 (0) 1 4345 2256
agence@champsaur.com
www.champsaur.com

298 Alison Vance
Jeffrey's Interiors
8 North West Circus Place
Edinburgh EH3 6ST
Tel: +44 (0)131 247 8010
Fax: +44 (0)845 8822 656
alison@jeffreys-interiors.co.uk
www.jeffreys-interiors.co.uk

304 Kira Chuveleva
Kira Chuveleva
Varshavskoe Shosse
144/2, app 665
117519 Russia, Moscow
Tel: +791 040 500 02
kirachoo@gmail.com
www.chuveleva.ru

308 Alla Shumeyko
Alla Décor Studio
E & A Kosmodemiansky str 7 / 3 - 261
125130 Moscow
Russia
Tel: +7 903 770 87 17
alla.decor@mail.ru
www.alladecor.ru

312 Linda Steen
AS Scenario Interiorarkitekter
Pilestredet 75C, Inngang 48
NO354 Oslo, Norway
Tel: +47 928 93 000
ls@scenario.no
www.scenario.no

316 Kris Lin
KLID (Kris Lin Interior Design)
301 Room, The Fourth Building
No 1163 Hong Qiao Road
Changning District
Shanghai, China
Tel: +86 137 018 562 55
Fax: +86 216 20 99 918
kl_iad@vip.163.com
www.krislin.com.cn

320 Lesley Zaal
Etcetera Living
P.O. Box 75307, Beach Road
Jumeirah No. 1, Dubai, U.A.E.
Tel: +9 714 344 8868
Fax: +9 714 344 6884
info@etceteraliving.com
www.etceteraliving.com

324 Lígia Casanova
Atelier Lígia Casanova
LX Factory, Rua Rodrigues Faria, 103
1350-501 Lisbon, Portugal
Tel: +351 213 955 630
atelierligiacasanova@gmail.com
www.ligiacasanova.com

330 Jimmie Karlsson, Martin Nihlmar, Sally Anne McCoy
Jimmie Martin & McCoy
77 Kensington Church Street
London W8 4BG
Tel: +44 (0)207 937 7785
Fax: +44 (0)7092 881 218
contact@jimmiemartinandmccoy.com
www.jimmiemartinandmccoy.com

334 Michael Attenborough
Radisson Edwardian Hotels
140 Bath Road, Hayes
Middlesex UB3 5AW
Tel: +44 (0)207 629 7777
Fax:+44 (0)207 629 1459
attenbom@raddison.com
www.raddisonblu-edwardian.com

340 Roger Pop
aa studio
Calea Dorobantilor
nr 124 Sect. 1 ap. 6
Bucharest
Romania
Tel: +407 233 65 843
office@aastudio.ro
www.aastudio.ro

346 Rosa May Sampaio
Rosa May Decoracao
Rua Alemanha 691 - Jd Paulistano
São Paulo 01448 010
Brazil
Tel: +55 11 3085 7100
rosamaysampaio@terra.com.br
www.rosamaysampaio.com.br

350 Christian Baumann
Abraxas Interieur
Hegibachstrasse 112
8032 Zurich
Switzerland
Tel: +41 44 392 21 92
Fax: +41 44 392 21 93
info@abraxas-interieur.ch
www.abraxas-interieur.ch

356 Steve Leung
Steve Leung Designers
30/F Manhattan Place
23 Wang Tai Road
Kowloon Bay
Kowloon
Hong Kong
Tel: +852 2527 1600
Fax: +852 2527 2071
sld@steveleung.com
www.steveleung.com

364 Daniel Kostiuc & team
Intarya
8 Albion Riverside
8 Hester Road
London SW11 4AX
Tel: +44 (0)207 349 8020
Fax: +44 (0)207 349 8021
info@intarya.com
www.intarya.com

370 Janine Lazard
Janine Lazard Interiors
PO Box 412254
Craighall 2024
Johannesburg
South Africa
Tel: +27 11 784 0090
Fax:+27 11 784 8818
lazards@mweb.co.za
www.janinelazard.com

374 Ivan Cheng
Ivan C. Design Limited
Flat 1, 13 F Block A
Gold Way Industrial Centre
16-20 Wing Kin Road
Kwai Chung
N.T. Hong Kong
Tel: +852 987 37 703
Fax: +852 255 620 23
icdl20111@live.com
www.icdl-hk.com

380 Joanna Berryman
Matrushka
4 Willoughby Road
London NW3 1SA
Tel: +44 (0)7435 1386
Mobile: +44 (0)7796 002344
info@matrushka.co.uk
www.matrushka.co.uk

386 Chang Ching - Ping
Tien Fun Interior Planning
12F-5 No 211 Chung Min Road
Taichung
Taiwan
Tel: +886 4 220 18 908
Fax: +886 4 220 36 910
tien.fun@msa.hinet.net

392 Olga Stupenko
OS Empire of Design
Bolshoy Chudov Lane 5
Moscow 119270
Russia
Tel: +7 (485) 773 6440
os@olgastupenko.com
www.olgastupenko.com

396 Anemone Wille Våge
Anemone Wille Våge
Bygdoy Alle 58B
0265 Oslo
Norway
Tel: +47 22 55 32 00
alexandra@anemone.no
www.anemone.no

402 Joseph Sy
Joseph Sy & Associates Limited
17/F Heng Shan Centre
141-145 Queen's Road East
Wan Chai
Hong Kong
Tel: +852 28 66 13 33
Fax: +852 28 66 12 22
design@jsahk.com
www.jsahk.com

406 Kari Arendsen
Intimate Living Interiors
143 South Cedros Avenue
Suite C203
Solana Beach
CA 92075
U.S.A.
Tel: +858 436 7127
Fax: +858 436 7140
info@intimatelivinginteriors.com
www.intimatelivinginteriors.com

412 Ichiro Shiomi, Etsuko Yamamoto
Spinoff Co., Ltd
6F 2-8-2 Ebisuminami
Shibuya-ku
Tokyo 150-0022
Japan
Tel: +81 3 5725 5171
Fax: +81 3 5725 5172
info@spinoff.cc
www.spinoff.cc

416 John G. Rufenacht
John Rufenacht
5013 Wyandotte St.
1-North
Kansas City
Missouri 64112
USA
Tel: +1 816 561 7795
rufenacht@sbcglobal.net
www.rufenachtinteriors.com

422 Thomas Dariel
Dariel Studio
3/F Building 1
Lane no 12 N
1384 Wanhangdu Road
Near Huayang Lu
200042 Shanghai
China
Tel: +86 (0) 21 6191 0102
Fax: +86 (0) 21 6191 0103
Mobile: +86 138 166 244 34
dariel@lime388.com
www.darielstudio.com

428 Robert Mills
Robert Mills Architects
1st Floor
10 Grattan Street
Prahran 3181
Australia
Tel: +03 9525 2406
info@robmills.com.au
www.robmills.com.au

434 Fabio Galeazzo
Galeazzo Design
Rua Antonio Bicudo 83
Sao Paulo
SP Brazil
Tel: +55 (11) 3064 5306
info@fabiogaleazzo.com.br
www.fabiogaleazzo.com.br

438 Tang Chi-Li, Yeh Chun-Erh, Law Louis
TF Designs Global
No 2. Lane 39
Liyuan Road
Dali District
Taichung City 412
Taiwan, R.O.C.
Tel: +886 4 220 35 358
Fax: +886 4 220 26541
tfdg01@yahoo.com

442 Sonia Korn
Sonia Korn Ltd
310 East 53rd St.
NYC 10022
NYC
U.S.A
Tel: +1 212 308 3308
Mobile: +1 917 971 6792
sonia@soniakorn.com
www.soniakorn.com

446 Ryu Kosaka
A.N.D.
Qiz Aoyama 3F 3-39-5 Jingumae
Shibuya-Ku
Tokyo
Japan 150 - 0001
Tel: +81 3 541 26785
Fax +81 3 3479 8050
madoka_sakano@npmurakougei.co.jp

450 Kathryn Ireland
Kathryn M. Ireland
9443 Venice Blvd., Suite B
Culver City
CA 90232, USA
Tel: +310 837 8890
Fax: +310 837 8898
info@kathrynireland.com
www.kathrynireland.com

456 Staffan Tollgård
Staffan Tollgård
100 Westbourne Studios
242 Acklam Road
London W10 5JJ
Tel: +44 (0)207 575 3185
Fax: +44 (0)207 575 3186
staffan@tollgard.co.uk
www.tollgard.co.uk

460 Peter Phan
Peter Phan Design Consultancy Ltd
First Floor Studio
8 Broadstone Place
London W1U 7EP
Tel: +44 (0)207 535 5160
Fax: +44 (0)207 486 3551
peter@peterphandesign.com
www.peterphandesign.com

468 Angelos Angelopoulos
Angelos Angelopoulos Associates
5 Proairesiou Street
Athens, 11636
Greece
Tel: +30 210 756 7191
design@angelosangelopoulos.com
www.angelosangelopoulos.com

472 Greg Natale
Greg Natale Design
Studio 6, Level 3
35 Buckingham Street
Surry Hills NSW 2010
Australia
Tel: +61 (0) 2 8399 2103
Fax: +61 (0) 2 8399 3104
info@gregnatale.com
www.gregnatale.com

476 Yvonne O'Brien
The Private House Co
PO Box 3277 Dainfern
2055 Johannesburg
South Africa
Tel: +27 114 65 56 00
Fax: +27 114 65 7900
yvonneo@icon.co.za

482 Philip Tang & Brian Ip
P Tang Studio Ltd
Rm 5-6
4/F Hopeful Factory Centre
10-16 Wo Shing Street, Fo Tan
Hong Kong
Tel: +852 2669 1577
Fax: +852 2669 3577
office@ptangstudio.com
www.ptangstudio.com

486 Carmo Aranha and Rosário Tello
Sa, Aranha & Vasconcelos
Rua Vale Formoso No 45
1950-279, Lisbon
Portugal
Tel: +351 218 453 070
Fax: +351 218 495 325
info@saaranhavasconcelos.pt
www.saaranhavasconcelos.pt

494 Chris Dezille
Honky
Unit 1, Pavement Studios
40-48 Bromells Road
London SW4 OBG
Tel: +44 (0)207 622 7144
Fax: +44 (0)207 622 7155
info@honky.co.uk
www.honky.co.uk

500 Liv Norun
Slettvoll Mobler AS
Skaffarvegen 105, 6200 Stranda
Norway
Tel: +47 70 26 88 10
Fax: +47 70 26 88 40
slettvoll@slettvoll.no
www.slettvoll.no

504 Zeynep Fadillioglu
ZF Design
A. Adnan Saygun Cad Koru Sokak
Ulus 1, Block D 3
Ulus
Istanbul
Tel: +90 (212) 287 09 36
Fax: +90 (212) 287 09 94
design@zfdesign.com
www.zfdesign.com

鸣 谢

本书作者及出版社向所有为本书提供项目资料的设计师和所有者表示诚挚的感谢；同时，感谢以下摄影师：

Ruth Cincotta, Sean Finnigan, Peter Rymwid, H. Durston Saylor, Robert Brantley, Jaideep Oberoi, Deidi Von Schaewen, Renaud Marin, Myriam Ramel, Billy Cunningham, Eric Piasecki, Scott Francis, Gordon Beall, Marina Faust, Ken Hayden, Eric Laignel, Ron Reeves, Jenni Hare, Eloise Fotheringham, Blandon Belushin, Courtesy of The Cosmopolitan of Las Vegas, Ellen Silverman, Frank Oudeman, Paul Warchol, Nikolas Koenig, Didier Delmas, Francis Amiand, Tony Figueira, Ajax Law Ling Kit, Virginia Lung, Louis Leung, The Pig Hotel, Matthew Millman, Traver Rains, Dook, David Glomb, Wu Yongcheng, Zhou Yuedong, Robert Miller, Francisco Almeida Dias, Edward Hendricks, Frederic Ducout, Wu Yongchang, Sean Myers, Will Clarkson, Aude Benoit, Bamba Sourang, Christian Bossu-Picat, Christophe Bielsa, Marcel Jolibois, Christian Rochat, Elena Akimova's Photographers, Ted Yarwood, Emily Gilbert, Maggie Ma, Franciso Almeida Dias, Anthony Tahlier, Eric Hausman, Nathan Kirkman, Jianguo Xie, Caroline Ryan, Tim McGhie, Katie Bateman, Steve Hall, Wim Pauwels, Pere Planells, Marcel Joliebois, Tom Sullam, Peter Bennett, Simon Winson, Patrick Steel, Diamond Dining, Nacasa & Partners, Masaya Yoshimura, Five Season, Giorgio Baroni, Antoine Bootz, Marili Forastieri, Michael Moran, Jason Penney, Durston Saylor, Samsam Siew Shien, Kyle Yu, Rainer Hofmann, Therese-Studer-Strasse, Joao Morgado, Philip Vile, Zinur Razuldinov, Wentao She, Tom Mannion, Valentino Fialdini, Beto Riginik, Romulo Fialdini, Daniel Hase, Andrew Elderfield, Will Webster, Giorgio Baroni, Thierry Cardineau, Morten Andersen, Stian Broch, Fritz von der Schulenburg, Paulius Gasiunas, Bernard Touillon, Neale Smith, George Ferguson, Paul Blackmore, Sergey Ananiev, Mikhail Stepanov, Olga Ludvig, Gatis Rozenfelds, Zhou Yaodong, Neil Corder, Oliver Jackson, Bruno Barbosa, Manuel Gomes da Costa, Rick Schultz, Ben Knight, Adrian Houston, Tim Winter, Chris Tubbs, Cornel Lazia, James Stewart Granger, Marc Zwicky, Mr Virgile Simon Bertrand, Mr Ulso Tsang, Mr Chen Wei Zhong, Richard Waite, David Garcia, Tanapol Kaewpring, Mel Yates, Cheung Chi Wai, Ivan Cheng, Zhi Kang, Peter Ying, Katarina Malmstrom, Javier Garcia-Alzorriz, Ian Andrew Martin, Natalie Dinham, Leo Bieber, Simon Williams, Liu Chun-Cjieh, Kirill Ovchinnikov, Elisabeth Aarhus & Studio Dreyer Hensleys, Joseph Sy, Brevin Blach photography, Nacasa & Partners inc, Sinichi Sato, Peter Vitale, Derryck Menere, Shannon McGrath, Earl Carter, Richard Powers, Marco Antonio, Celia Weiss, Evelyn Muller, Jay Lin, Manel Moniz, Ben Wood, Goncalo Fabiao, Takeshi Nakasa, Victoria Pearson, Tim Beddows, Jonathan Beckerman, Kathryn Ireland, Mary Wadsworth, Richard Gooding, Daniella Cesarei, Vasco Celio, Robert Hanson, Mitropoulos Costas, Anson Smart, Elsa Young, Ptang Studio Ltd, Marc Paris, Francisco de Almeida Dias, Carlos Eduardo Bleck de Vasconcellos e Sa, Simon Winson, Margaret M. De Lange, Roar Nord Jensen, Eirik Slyngstad, Koray Erkaya, Benno Thoma.

图书在版编目（CIP）数据

室内设计奥斯卡奖/（英）沃勒编著；叶玮译. --
南昌：江西美术出版社，2012.12
ISBN 978-7-5480-1762-2

Ⅰ.①室… Ⅱ.①沃… ②叶… Ⅲ.①室内装饰设计
－作品集－世界－现代 Ⅳ.①TU238
中国版本图书馆CIP数据核字(2012)第268110号

Andrew Martin—Interior Design Review Volume 16
Copyright © Andrew Martin International
合同登记号：124-2012-508

《FRAME国际中文版》杂志社 策划
地　址：北京市朝阳区酒仙桥路4号，798艺术区，751时尚设计广场，悠之设计中心二层
电　话：010-8459 9571 / 9572
企　划：北京江美长风文化传播有限公司
地　址：北京市海淀区花园路甲13号院7号楼庚坊国际10层
电　话：010-8229 3750

出 品 人：陈　政
责任编辑：王国栋
编辑助理：楚天顺
特约编辑：叶　玮
封面设计：刘一霖
版式设计：刘一霖

室内设计奥斯卡奖

（英）马丁·沃勒　编著
叶玮　译

出版发行：江西美术出版社
地　　址：南昌市子安路66号 江美大厦（邮编330025）
网　　址：http://www.jxfinearts.com
经　　销：全国新华书店
印　　刷：上海锦良印刷厂
开　　本：665×980　1/8
印　　张：64
字　　数：512千字
版　　次：2012年12月第1版
印　　次：2012年12月第1次印刷
书　　号：ISBN 978-7-5480-1762-2
定　　价：468.00元

本书由江西美术出版社出版，未经出版者书面许可，不得以任何方式抄袭、复制或节录本书的任何部分
版权所有，侵权必究
本书法律顾问：江西豫章律师事务所 晏辉律师
赣版权登字-06-2012-829